Guidelines for the Use of the Semipermeable Membrane Device (SPMD) and the Polar Organic Chemical Integrative Sampler (POCIS) in Environmental Monitoring Studies

By David A. Alvarez

Techniques and Methods 1–D4

U.S. Department of the Interior
U.S. Geological Survey

U.S. Department of the Interior
KEN SALAZAR, Secretary

U.S. Geological Survey
Marcia K. McNutt, Director

U.S. Geological Survey, Reston, Virginia, 2010

For more information on the USGS—the Federal source for science about the Earth, its natural and living resources, natural hazards, and the environment, visit http://www.usgs.gov or call 1-888-ASK-USGS

For an overview of USGS information products, including maps, imagery, and publications, visit http://www.usgs.gov/pubprod

To order this and other USGS information products, visit http://store.usgs.gov

Suggested citation:
Alvarez, D.A., 2010, Guidelines for the use of the semipermeable membrane device (SPMD) and the polar organic chemical integrative sampler (POCIS) in environmental monitoring studies: U.S. Geological Survey, Techniques and Methods 1–D4, 28 p.

Contents

Tables

Figures

Conversion Factors

Multiply	By	To obtain
Length		
centimeter (cm)	0.3937	inch (in.)
Volume		
microliter (µL)	0.0000338	ounce, fluid (fl. oz)
nanoliter (nL)	0.0000000338	ounce, fluid (fl. oz)
milliliter (mL)	0.0338	ounce, fluid (fl. oz)
liter (L)	0.2642	gallon (gal)
Mass		
kilogram (kg)	2.205	pound avoirdupois (lb)

Temperature in degrees Celsius (°C) may be converted to degrees Fahrenheit (°F) as follows:

$$°F=(1.8×°C)+32$$

Temperature in degrees Fahrenheit (°F) may be converted to degrees Celsius (°C) as follows:

$$°C=(°F-32)/1.8$$

Guidelines for the Use of the Semipermeable Membrane Device (SPMD) and the Polar Organic Chemical Integrative Sampler (POCIS) in Environmental Monitoring Studies

By David A. Alvarez

1. Introduction

The success of an environmental monitoring study using passive samplers, or any sampling method, begins in the office or laboratory. Regardless of the specific methods used, the general steps include the formulation of a sampling plan, training of personnel, performing the field (sampling) work, processing the collected samples to recover chemicals of interest, analysis of the enriched extracts, and interpretation of the data. Each of these areas will be discussed in the following sections with emphasis on specific considerations with the use of passive samplers.

Water is an extremely heterogeneous matrix both spatially and temporally (Keith, 1991). The mixing and distribution of dissolved organic chemicals in a water body are controlled by the hydrodynamics of the water, the sorption partition coefficients of the chemicals, and the amount of organic matter (suspended sediments, colloids, and dissolved organic carbon) present. In lakes and oceans, stratification because of changes in temperature, water movement, and water composition can occur resulting in dramatic changes in chemical concentrations with depth (Keith, 1991). Additional complications related to episodic events, such as surface run-off, spills, and other point source contamination, can result in isolated or short-lived pulses of contaminants in the water.

The application of passive sampling technologies for the monitoring of legacy and emerging organic chemicals in the environment is becoming widely accepted worldwide. The primary use of passive sampling methods for environmental studies is in the area of surface-water monitoring; however, these techniques have been applied to air and groundwater monitoring studies. Although these samplers have no mechanical or moving parts, electrical or fuel needs which require regular monitoring, there are still considerations that need to be understood in order to have a successful study.

Two of the most commonly used passive samplers for organic contaminants are the semipermeable membrane device (SPMD) and the polar organic chemical integrative sampler (POCIS). The tips given in this document focus on these two samplers but are applicable to most types of passive sampling devices. The information in this guide is heavily weighted towards the sampling of water; however, information specific to the use of SPMDs for air sampling will also be covered.

2. Before Heading to the Field

The success of a study using passive samplers, or any sampling method, begins in the office or laboratory with the formulation of a sampling plan and training of personnel. The sampling plan should include the goals of the study, selection of target chemicals and laboratories capable of performing the work, and identification of quality control measures. If possible, a reconnaissance trip to the study sites should be made before the fieldwork so the best deployment plan and set-up can be devised. If a reconnaissance trip cannot be made, be prepared to use multiple deployment options. Suggestions for securing the samplers in the field are described in Section 3.

The average timespan of a field deployment is 30 days. The actual time in the field is not important provided it is known how long the sampling devices were in the water. Deployments from one week to one year have been performed. Short deployments will result in smaller volumes of water being sampled, thereby limiting some of the advantages of using a passive sampler. Long deployments can result in changes in the sampling kinetics from an integrative sampler to an equilibrium sampler for certain contaminants and a substantial buildup of a biofilm that could inhibit the ability of the sampler to accumulate chemicals. Long deployments also pose a greater risk of damage or loss because of high water events in streams, and vandalism. Short and long deployments are arbitrary descriptive terms as the actual lengths can vary greatly depending on the chemical targeted and environmental variables. Generally, field deployments are limited from 2 to 3 months.

2a. What type and how many samplers do you need?

The decision on which type of passive sampler to use is dependent on the chemicals targeted for the study. Often times, SPMDs and POCIS are used together to obtain a sample more representative of the entire range of organic contaminants than can be obtained with a single sampler (Petty and others, 2004). Although there is some overlap between the devices with regards to what chemicals can be sampled, the following guidelines can help determine which type of sampler may work best for the targeted chemicals.

SPMDs are generally used for sampling neutral organic chemicals with a log octanol-water partition coefficient (K_{ow}) greater than 3. Polycyclic aromatic hydrocarbons (PAHs), polychlorinated biphenyls (PCBs), chlorinated pesticides, polybrominated diphenyl ethers (PBDEs), dioxins, and furans are all commonly measured using SPMDs. Hydrophobic contaminants often related to wastewater effluents such as fragrances, triclosan, and phthalates are also often detected with field deployed SPMDs.

SPMDs can be made at various lengths to suit specific applications of a study. The user should be aware that the amount of chemical sampled is related to the surface area of the sampling device; therefore, using smaller SPMDs decreases the amount of chemical sampled. In certain cases, such as screening studies in suspected highly polluted areas, smaller versions of the SPMD can be used (Goodbred and others, 2009). Two versions of the SPMD are commercially available depending on the purity of the triolein used in their manufacture. The standard SPMD contains 99 percent pure triolein; however, this grade of triolein contains residual methyl oleate and oleic acid from the triolein synthesis which may be coextracted and interfere with some instrumental and bioassay procedures unless steps are implemented to remove it during extract processing (Petty and others, 2000). The ultra-high purity SPMD contains triolein which has undergone an additional purification step in the laboratory to essentially remove all traces of oleic acid and substantially reduce the amount of methyl oleate in the final extract (Lebo and others, 2004). It is the recommendation of this author to only use the ultra-high purity SPMDs as it increases the quality of the SPMD and greatly simplifies the analyses at a nominal additional cost.

POCIS are designed to sample the more water soluble organic chemicals with log K_{ow}s less than (<) 3. This includes most pharmaceuticals, illicit drugs, polar pesticides, phosphate flame retardants, surfactants, metabolites and degradation products. Although log K_{ow}s for steroidal hormones, fragrances, triclosan, and other chemicals related to wastewater effluents are generally greater than (>) 3, these compounds are often preferentially sampled by the POCIS (fig. 1).

There are two configurations of the POCIS that are commercially available. These are referred to as the pesticide-POCIS and the pharmaceutical-POCIS. These names are misleading as each type can sample a wider range of chemicals than the names suggest. The difference in the POCIS

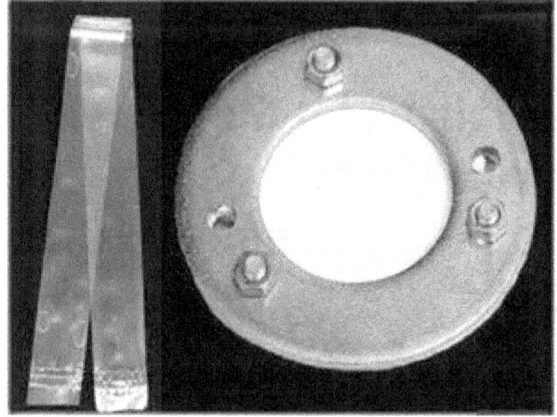

Not to scale

Figure 1. Semipermeable membrane device (SPMD, left) and the polar organic chemical integrative sampler (POCIS, right). Photographs by Randal Clark, U.S. Geological Survey (left) and Environmental Sampling Technologies (right).

configurations is the solid sorbent, the component that traps the sampled chemicals, used in the POCIS construction. The pesticide-POCIS uses a triphasic admixture of Isolute® ENV+ and Ambersorb® 1500 or 572 carbon dispersed on S-X3 BioBeads®. This configuration was originally designed for sampling pesticides and hormones; however, as more chemicals were added to the targeted chemical lists, it was determined that many chemicals, such as pharmaceuticals with multiple functional groups, were difficult to recover from the sorbent. For that reason, a second POCIS type, the pharmaceutical-POCIS, was developed containing the sorbent Oasis HLB. Oasis HLB is typically considered a universal sorbent in environmental analyses and has been used to extract a wide assortment of chemical classes from water. There also is the advantage with the Oasis HLB of numerous published methods for the solvent combinations for use in recovering chemicals from the sorbent, potentially simplifying method development (Reemtsma and Quintana, 2006; Buchberger, 2007; Waters, 2008; Pacáková and others, 2009). Vermeirssen and others (2005) determined that the pesticide-POCIS had a higher efficiency for accumulation of some chemicals than the pharmaceutical-POCIS and may be advantageous in specific applications. However, it is this author's recommendation to use only the pharmaceutical-POCIS when targeting a wide range of chemicals unless the users have prior knowledge that targeted chemicals can be adequately sampled and recovered by the pesticide-POCIS configuration.

The analytical requirements of the study will dictate the number of passive samplers needed. The extract from a single passive sampler or a composite of extracts from multiple samplers is often needed for a particular analysis (dialysates from four to six SPMDs are often combined for dioxin analysis). Occasionally, the extract from a single passive sampler can be split between multiple analyses. Knowledge of the mass of a chemical, total number of nanograms (ng) for example, which must be sampled to meet the detection criteria of the chemical analysis or toxicity tests will affect the study design. The number of samplers needed can be estimated using the following equation (eq.):

$$R_s\, t\, n\, C_c\, P_r\, E_t > MQL\, V_i \qquad (1)$$

where C_c is the predicted environmental chemical concentration,

   t is the deployment time in days,

   R_s is sampling rate in liters of water extracted by the passive sampler per day (L/d),

   P_r is the overall method recovery for the analyte (expressed as a factor of one; therefore 0.9 is used for 90 percent recovery),

   n is the number of passive samplers combined into a single sample,

   E_t is the fraction of the total sample extract which is injected into the instrument for quantitation (0.001 if 1 microliter, µL, of a 1 milliliter, mL, sample is injected),

   MQL is the method quantitation limit, and

   V_i is the volume of standard injection (commonly 1 µL).

It may be necessary to contact the analytical laboratory for many of these variables. If eq. 1 is true, then the number of samplers selected (n) is suitable. If the result of the left side of the equation is less than the result of the right side, then n should be increased. As an example, assume you are using a single SPMD ($n=1$) for the determination of pyrene in a 30-day deployment ($t=30$) and the desired concentration level to be measured is (C_c) is 1 ng of chemical per L of water. Pyrene has an approximate R_s of 5 L/d and a P_r of 85 percent. The instrumental method has a MQL of 10 picograms/microliter (pg/µL), V_i of 1 µL and the final sample volume is 1 mL ($E_t=0.001$). Using eq. 1, 128 pg > 10 pg, therefore the relation is true and a single SPMD is sufficient to meet the needs of the study.

In addition to the chemical analysis of these samplers, the SPMD and POCIS have been combined with an array of *in vitro* and *in vivo* bioindicator tests to determine the potential effect on biota from exposure to the complex mixtures of chemicals present at a site. SPMDs have a long history of use as a surrogate biological organism as they mimic the accumulation of dissolved chemicals via the respiration of an organism (Huckins and others, 2006). SPMDs have been used in combination with a host of bioindicator and classic toxicity tests including, but not limited to, Microtox, Mutatox, mixed function oxygenase (MFO) induction-ethoxyresorufin-*O*-deethylase (EROD) activity, sister chromatid exchange, vitellogenin induction, enzyme-linked immunosorbent assay (ELISA), Daphtoxkit F, Ames mutagenicity tests, the yeast estrogen screen (YES), as well as whole organism dosing and exposure studies (Huckins and others, 2006; Rastall and others, 2004). The POCIS has been primarily used in combination with the YES assay, but would be amenable to other

bioindicator tests as well (Alvarez and others, 2007, 2008a; Vermeirssen and others, 2005).

At the time of this writing (2010), the SPMD is covered by two U.S. government patents and one Canadian patent (Huckins and others, 1992, 1995, 1996) and POCIS is covered by one U.S. government patent (Petty and others, 2002). The sole commercial vendor in North America is Environmental Sampling Technologies, Inc. (EST Labs) in St. Joseph, Missouri. The inventors of these devices at the U.S. Geological Survey (USGS), Columbia Environmental Research Center (CERC) also maintain the right to construct, process, and use these samplers as part of their research projects and collaborations. Contact EST Labs or CERC to determine the availability of the samplers.

2b. Performance Reference Compounds (PRCs)

Environmental factors such as water flow, temperature, and the buildup of a biofilm on the sampler's surface can affect the rate at which chemicals are sampled. These factors are site-specific and can greatly reduce the accuracy of the estimated water concentrations. In an effort to account for these effects, the performance reference compound (PRC) approach was developed (Huckins and others, 2002a). A PRC is a chemical with moderate to high fugacity (escaping tendency) that is added to the sampler during fabrication. By measuring the amount of PRC loss during deployment in the field, adjustments to the theoretical or experimentally derived sampling rates of targeted chemicals can be made to reflect the site-specific sampling rates. The mathematical use of PRC data will be discussed in Section 5a.

Proper selection of a PRC requires knowledge of the possible occurrence of a PRC in the environment, the predicted rate of loss from the sampler, and the analytical capabilities of the laboratory. PRCs must not occur naturally as amounts accumulated from the environment will bias the PRC loss measurement. Ideally, deuterated or carbon-13 (^{13}C)-labeled versions of targeted chemicals are selected. In cases where labeled chemicals are not available or are cost-prohibitive, non-labeled chemicals can be used. For example, the PCB congeners 14, 29, and 50 are often used as PRCs since they do not occur in the environment. Huckins and others (2006) determined that it is not necessary to have a PRC for each chemical class. Using the current models for determining PRC-derived sampling rates, a PRC can be used to calculate site specific sampling rates with increased accuracy through a range of chemical classes and log K_{ow}s.

In addition to adding PRCs, it is common to add a photolysis surrogate, a chemical that can degrade upon exposure to sunlight but has a low fugacity, therefore, it will not be lost because of diffusion out of the SPMD during the field exposure. Dibenz[a,h]anthracene-d_{14} is a PAH commonly used for this application. Like many PAHs and some other chemicals, it is sensitive to exposure to sunlight in the ultraviolet A and B bands (UVA and UVB). The low-density polyethylene membrane of the SPMD is transparent to UVA and UVB; therefore, photosensitive chemicals can degrade in clear waters. This can occur not only in shallow streams, but also at deeper depths, such as the clear waters of the Caribbean Sea (Garrison and others, 2005; Bartkow and others, 2007).

For PRC data to be useful, chemicals which have a loss of 20 to 80 percent from the SPMD should be selected as the measurements should be significantly different than the analytical variability of the method (Huckins and others, 2002a). The rate of PRC loss increases as the chemical's log K_{ow} decreases. Increased exposure temperatures and flow (water or air) across the membrane surface also increase the PRC loss rates. It is advantageous to add multiple PRCs to ensure at least one PRC will have a suitable loss for computation. Selection of a PRC with a log K_{ow} greater than 5.5 to 6 should be avoided, as the amount of the PRC lost during exposure will be very low except at high temperatures or prolonged field deployments.

Finally, communication with the analytical laboratory regarding the selection of the PRCs must occur before use. The laboratory needs to be able to analyze the PRCs with existing methods or be willing to modify methods, often at additional costs, to accommodate the PRCs. The selection of the PRCs also must not interfere with other chemicals in the targeted analysis. Some chemicals chosen as PRCs, such as PCB congener 29, are often used by some laboratories as internal standards and, therefore, cannot be used.

The PRC approach has been well defined for SPMDs; however, its use with the POCIS is much less uncertain. Implicit in the PRC approach for passive samplers is the assumption that the overall uptake and release of targeted chemicals and PRCs are governed by first-order kinetics and that the sum of resistances to mass transfer across sampler associated barriers is equal in both directions (Alvarez and others, 2007). This assumption may not be valid for solid-phase extraction sorbents, like those used in the POCIS, because of the fundamental differences between chemical partitioning and adsorption phenomena. In these cases, the surrounding water is not a strong enough solvent to elute (remove) the target chemicals from the solid support. Mazzella and others (2007; 2010) have determined that a highly polar pesticide degradation product, desisopropyl atrazine, has potential as a PRC for POCIS and has improved accuracy of water concentration estimates for pesticides with a similar structure. It is unknown if this potential PRC can be applied universally to other target chemicals with vastly different molecular structures and properties.

2c. Quality Control

There are many types of quality control (QC) measures which can be part of a laboratory's good laboratory practices (GLP) plan. For the purposes of this document, only the types of blanks and spikes recommended for use in passive sampling studies are discussed. Many of these QC sample types are common with all environmental studies regardless of the

sample matrix. In addition to the types of QC samples listed below, surrogate standards (stable isotope labeled versions of the target chemicals) may be added to the passive samplers before or immediately following the initial dialysis or extraction. These surrogates can be used to correct for procedural losses (percent recovery) and potential interferences.

With the number of potential QC sample types used, it is common for the number of QC samples required for a study to range between 10 and 50 percent of the total number of samples. The exact percentage of QC samples needed is dependent on the project objectives determined during the planning stage. It is not always economically feasible to include a large QC sample set with every study, but not doing so potentially compromises the interpretation of the results. At a minimum, the following types of QC samples should be used: fabrication blanks, field blanks, and matrix spikes. Occasionally, a field blank can be exposed to the air at multiple sites resulting in a cumulative field blank. Assuming that no contamination is measured in the cumulative field blank, a cumulative field blank is acceptable. If contamination is detected, it is impossible to determine at which site the contamination occurred, therefore, all field deployed samples covered by the cumulative field blank would be biased by the blank result.

Blanks

The types of blanks used with passive samplers can include Fabrication, Field, Trip, and Laboratory Blanks. In addition, reagent blanks and instrument or procedure-specific blanks also can be used.

Fabrication Blanks

Fabrication blanks occasionally are referred to as Day 0 (zero) blanks. They are fabricated concurrently with the field deployed samplers and are stored under an inert atmosphere at <-20 degrees Celsius (°C) until they are processed along with the field samplers. Fabrication blanks account for interferences or contamination incurred from the SPMD or POCIS components, storage, processing, and analysis. If PRCs are used with the field samplers, the fabrication blanks also should be spiked with the PRCs.

Field Blank

Field blanks are stored in airtight containers and are transported to the field sites in insulated containers filled with blue ice or wet ice sealed in plastic bags. During the deployment and retrieval operations (the time the field passive samplers are exposed to air), the lids to the field blank containers are opened allowing exposure to the surrounding air. Field blanks account for contamination during transport to and from study sites, exposure to airborne contaminants during the deployment and retrieval periods, and from storage, processing and analysis. If using PRCs, the field blanks also should be spiked with the PRCs.

Trip Blanks

Trip blanks are often confused with field blanks, but there is one specific difference. Trip blanks are never exposed to the environment, they remain sealed in their shipping containers accounting for contamination during the transport to and from the study sites (Huckins and others, 2006; Keith, 1991). Trip blanks may be used with water sampling studies, but are generally used only with air sampling studies. If PRCs are used with the field samplers, the trip blanks also should be spiked with the PRCs.

Laboratory Blanks

Laboratory blanks are either SPMD or POCIS blanks and are manufactured at the time chemicals are extracted from the samplers. This type of blank can replace a reagent blank as it accounts for any contamination because of the processing and analysis of the samplers.

Spikes

Spikes are used to determine the recovery of targeted chemicals and potential interferences which may affect the analyses of the samplers. Spikes generally fall into two categories: matrix and procedural.

Matrix Spikes

Matrix spikes are SPMDs or POCIS prepared with a known quantity of targeted chemicals. This type of spike is carried throughout the whole processing scheme to determine the percent recovery of the targeted chemicals at the laboratory during analysis and to establish control limits for the analytical process. For some instrumental techniques such as liquid chromatography/mass spectrometry (LC/MS), a matrix spike can help determine if ion suppression of the target chemicals will occur because of the passive sampler matrix. This spike will not determine if ion suppression will occur from cosampled chemicals originating from the study site. If multiple chemical classes are to be targeted, separate matrix spikes are commonly prepared for each class or analytical method in order to simplify quantitation.

Procedural Spikes

Procedural spikes can be used to determine the recovery of target compounds for an individual procedural step, multiple steps, or for the whole analytical process. This type of spike is made of the test chemicals in a suitable solvent, but without the passive sampler matrix. In laboratories with appropriate licenses, radio-labeled compounds (^{14}C- or ^{3}H-labeled) are often used as procedural spikes to provide a rapid indication of the performance of an individual step.

2d. Limitations of using SPMDs and POCIS

The applications of using SPMDs and POCIS for environmental monitoring studies have been widely reported, however, there are some limitations to the technologies which should be considered before starting a passive sampling study. These limitations apply to all types of passive samplers, not just SPMDs and POCIS.

Passive samplers are designed to be long-term (weeks to months) integrative samplers. Generally, these samplers will provide little benefit over traditional discrete (grab) samples for study periods less than one week. Sampling devices which have a low capacity and quickly reach equilibrium, such as the solid-phase microextraction device (SPME), are better suited for short-term samplings.

An advantage of using an integrative sampler such as the SPMD and POCIS is that episodic events (surface runoff, spills, and other unpredictable sources of contamination) can be sampled without the cost and challenges of trying to catch the events with trained staff; however, because of the sampling nature of the devices, it is impossible to determine when the event occurred during the deployment period or know the maximum concentration of a chemical related to the event. Integrative samplers provide data as a time-weighted average concentration of a chemical within the whole exposure period.

As mentioned previously, it is important to discuss the study needs with the analytical laboratory before beginning fieldwork. Many analytical laboratories have not worked with passive samplers and may be uncertain of how to process the media or extract chemicals from it. Passive sampler extracts are often not as difficult to work with as samples of other environmental matrices and are compatible with common methods the laboratory may have in place. The reporting procedures of a laboratory should be discussed as many laboratories use automated reporting systems set up to report in the units of ng/L of water. These units are not suitable for a passive sampler extract as the desired units should be reported as total ng of chemical per SPMD or POCIS. The units of ng/SPMD or ng/POCIS are required for the calculations to estimate ambient water concentrations.

If data obtained from the passive samplers may be used for regulatory decisions, it should be determined if and how the passive sampler results will be accepted or admissible in court. In general, the passive sampling technologies for environmental monitoring are considered research methods by the regulatory community. A few states in the United States and a few European countries have started to accept passive samplers as a monitoring tool and the European Union has initiated studies to determine their potential acceptance (European Commission, 2009).

The POCIS is well-suited as a screening tool for determining the presence or absence of, sources, and relative amounts of chemicals at study sites. Estimation of ambient water concentrations requires knowledge of the sampling rate for each chemical measured. As of March 2010, POCIS sampling rates have been published for approximately 200 chemicals ranging from pesticides, surfactants, illicit drugs, to pharmaceuticals and personal care products (Alvarez, 1999; Alvarez and others, 2004; 2007; Arditsoglou and Vousta, 2008; Bueno and others, 2009; Harman and others, 2008; Li and others, 2010; MacLeod and others, 2007; Mazzella and others, 2007); however, within a period of a couple years, research on a global scale has increased and new sampling rates have been published aiding in the utility of the POCIS as a quantitative tool.

The PRC approach has been well defined for the SPMDs; however, this approach has only been demonstrated for the POCIS on a limited basis (Mazzalla and others, 2007; 2010). Implicit in the PRC approach is the assumption that the overall uptake and release rates of chemicals are governed by first-order kinetics and the sum of resistances to mass transfer are the same for the chemical flux into and out of the POCIS (Alvarez and others, 2007). This assumption may not be valid for solid phase sorbents, such as those used in the POCIS, with the exception of a limited number of chemicals because of fundamental differences between chemical partitioning and adsorption phenomena. Because of the strong sorptive properties of the adsorbents used in the POCIS, the POCIS may act as an infinite sink resulting in an anisotropic exchange (Huckins and others, 2006). Currently, the best alternative to having PRCs incorporated into a POCIS is to use an SPMD with PRCs as a surrogate as described by Alvarez and others (2007) and Harman and others (2008).

3. In the Field

Every site presents a different set of challenges. This section is meant to help you successfully use passive samplers in the field. A successful study involving passive samplers may require modifications to the deployment methods discussed later in this publication. Innovative ideas will solve problems encountered in the field. Also, following standard protocols for sample collection will minimize sample contamination.

3a. Equipment Needed

A variety of custom and commercially available deployment canisters are commonly used to protect the passive samplers in the field (fig. 2, table 1). The canisters are shipped to the field preloaded with the passive samplers with the exception of the commercially available canisters (fig. 2A and 2B). These canisters require that the SPMDs or POCIS (preloaded on racks) are inserted into the canister in the field. The canisters shown in figures 2A, 2B, and 2D are commercially available from EST Labs (St. Joseph, Missouri). The Prest-style canister (fig. 2C) was designed by Dr. Harry Prest, Santa Cruz, California, and is a custom design used by USGS. Two custom POCIS deployment canisters made from PVC components (figs. 2E and 2F) were also designed and used by USGS researchers.

Other custom deployment canisters can be constructed provided they meet certain criteria. The canisters must be durable to protect the passive samplers from damage and allow adequate water movement through the canister. Openings in the canister should be small enough to prevent large debris or organisms from entering the canister which may damage the passive samplers. Any materials used in the construction of the canisters must be free of any chemicals which may contaminate the passive samplers. For example, most metal parts, including stainless steel, come from the manufacturer with a coating of oils which serve as corrosion inhibitors and lubricants during manufacture. Plastic components may leach organic and inorganic chemicals that may be absorbed by the passive samplers. A thorough cleaning of the deployment canisters before loading with the passive samplers is critical. Cleaning methods may involve a dilute acid wash (to remove salts and loosen surficial sediments and biological growth), hot soapy water wash, tap or deionized water wash, and finally an organic solvent rinse starting with a polar solvent (isopropanol alcohol or acetone) followed by a nonpolar solvent (hexane).

The types of equipment needed for the deployment and retrieval of passive samplers can vary depending on the site and how the samplers are deployed. Some general equipment needs are listed below and in a checklist in the appendix. Water quality measurements, such as pH and specific conductance, are not needed for passive sampler measurements, but may be helpful in interpreting the data obtained from the passive samplers.

- Ice chest/cooler for transporting the passive samplers to/from the field
  - blue ice or wet ice (sealed in plastic bags)
  - canister(s) in sealed metal cans
  - trip/field blank(s)
- Tools
  - paint can opener
  - assorted tools (wrenches, pliers, cutters, saws)
  - thermometer
- Deployment hardware
  - cable, cable clamps, tie-down anchors, floats
  - signage, markings (personal choice because of potential vandalism)
- Field log book/sheets, digital camera

3b. Site Selection

Site selection should first be made based on the goals of the study. Because of accessibility of the site and logistics of

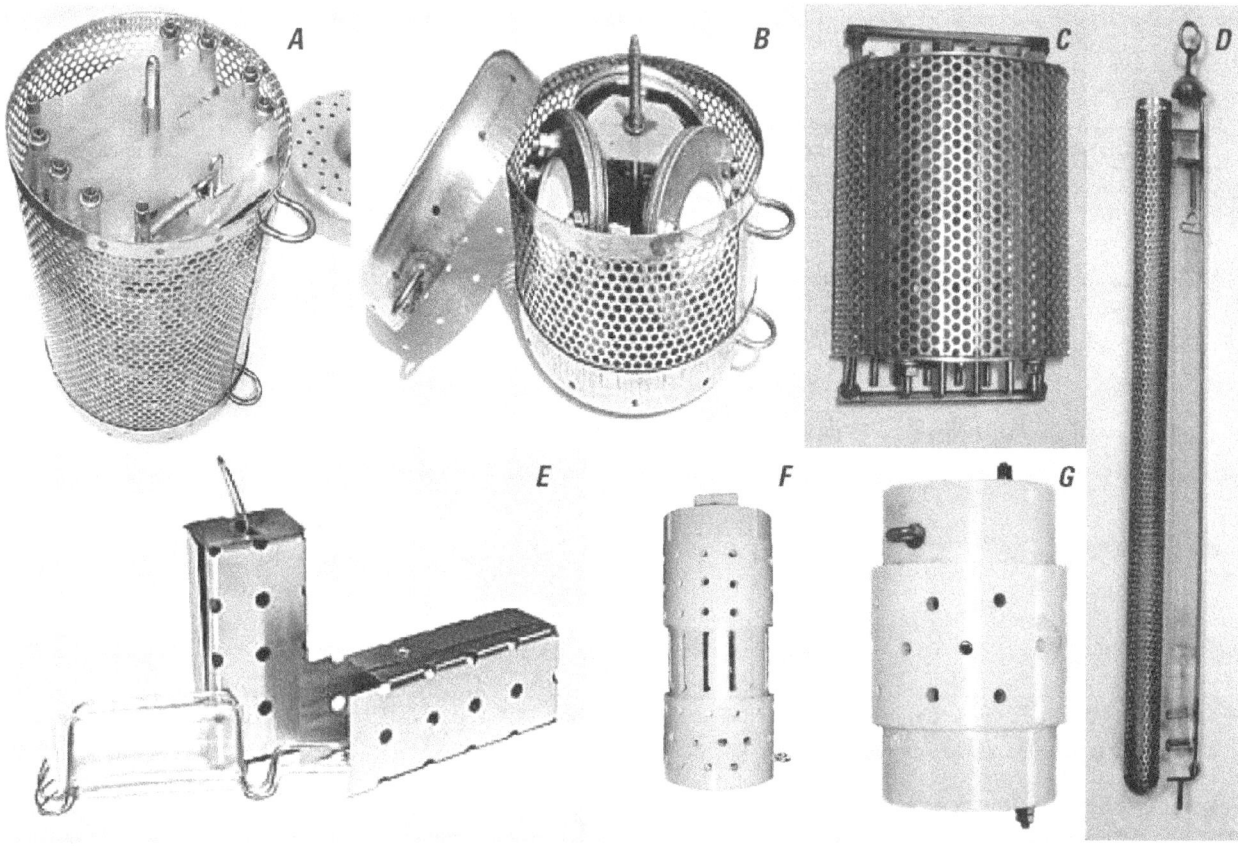

Not to scale

Figure 2. Custom and commercially available deployment canisters for semipermeable membrane devices (SPMDs) and polar organic chemical integrative samplers (POCIS). *A,* Commercially available large stainless steel canister for SPMDs or POCIS. *B,* Commercially available small stainless steel canister for SPMDs or POCIS. *C,* Custom "Prest-style" canister for SPMDs or POCIS. *D,* Commercially available stainless steel well sampling device. *E,* Commercially available stainless steel mini-SPMD canister. *F,* Custom large polyvinyl chloride (PVC) canister for POCIS. *G,* Custom PVC canister for POCIS. Photographs by Environmental Sampling Technologies *(A, B, E)* and David Alvarez, U.S. Geological Survey *(C, D, F, G).*

Available Water

The most important factor in the site selection is that the passive samplers remain submerged throughout the entire deployment period. If the samplers become exposed to air during the deployment, the possibility for contamination from airborne chemicals and loss of sampled chemicals to the surrounding air exists, in addition, estimation of ambient water concentrations will be impossible as it will not be known how many days the samplers were actually in the water. Water depth can be an issue with small streams and systems affected by tides. Small streams often do not have adequate depth to cover the deployment canisters (see dimensions in table 1) or may be prone to drying out because of lack of rain. This is a common situation when canisters are deployed during periods of high precipitation and left during periods of low

or no precipitation. Reconnaissance for site selection in small streams is more effective when the flow is near normal (base flow, not affected by recent rainfall). Selection of the most favorable sites is easier when the streams are at base flow. Allow for a minimum of 12 inches of water over the top of the deployment canister(s). Any less coverage can risk exposure to air if precipitation is less than average immediately before and during the deployment period.

If sufficient water depth is not available the options are limited. Finding a deep hole would be best or a hole can be dug in the streambed provided it is not too rocky. Digging holes can be problematic in streams with high suspended solids which could cause the canisters to become buried from sedimentation. Constructing temporary dams is not a favorable option because of unknown consequences upstream and downstream from the restricted water flow. More frequent site visits may be necessary to monitor water levels in shallow water streams. Terminating a deployment early is better than allowing the samplers to become exposed to the air, thereby invalidating data obtained from the samplers. If the study

Table 1. Dimensions of common SPMD and POCIS deployment canisters.

[SPMD, semipermeable membrane device; POCIS, polar organic chemical integrative sampler; EST, Environmental Sampling Technologies, commercial vendor; in, inch; lbs, pounds; PVC, polyvinyl chloride; ID, inner diameter]

Canister Type	Source	Dimensions	Weight	Capacity
Large stainless steel canister	EST	12 in x 6 ¼ in rings = 1 in x ¾ in	5 lbs (5 SPMDs) 9 lbs (6 POCIS)	5 standard[1] SPMDs or combination of SPMDs and POCIS
Small stainless steel canister	EST	6 in x 6 ¼ in rings = 1 in x ¾ in	3 lbs (2 SPMDs) 4.5 lbs (3 POCIS)	2 standard SPMDs or 1 rack of POCIS
"Prest-style" canister	Custom	5 ¼ in x 7 in handles = 4 ¾ in x 1 ¼ in	2.5 lbs (4 SPMDs)	4 standard SPMDs or combination of SPMDs and POCIS
Stainless steel well sampling device	EST	27 ½ in x 1 ¾ in	1.3 lbs (loaded)	2 standard SPMDs
Mini stainless steel SPMD canister	EST	5 in x 1 ¾ in	0.04 lb (loaded)	1 mini[2] SPMD
Large PVC canister	Custom	13 in x 5 in eye bolts = ½ in ID	3 lbs (empty) 9 lbs (loaded)	10 POCIS
Small PVC canister	Custom	8 ½ in x 5 in eye bolts = ½ in ID	2 lbs (empty) 7 lbs (loaded)	8 POCIS

[1] A standard SPMD is 91 centimeter (cm) long containing 1 mL of triolein.

[2] A mini SPMD is 15 cm long containing 0.17 mL of triolein.

area is affected by tides, the samplers must be deployed deep enough to remain submerged at low tide. Also, the deployment system used must allow for the retrieval of the samplers at high tide.

Flow

It is favorable to have the samplers in areas with flow as the volume of water sampled per day (sampling rate, R_s) is proportional to flow. However, areas with the highest flow should be avoided because of the potential for damage from floating debris and the difficulty in securing the samplers in a fixed location.

Physical Orientation

The physical orientation (horizontal, vertical, in line with flow) of the deployment canister in a water body generally is not an issue, provided the deployment canister has openings to allow for water exchange. Orientation may become an issue in streams or rivers with high levels of suspended solids. In these cases, it may be desirable to orientate the deployment canisters so the area of the canister with the fewest openings is facing upstream in order to minimize the amount of solids that may enter and become trapped inside the deployment canister. If suspended solids are predicted to be a problem, the deployment canister should be placed in the water behind some sort or obstacle to shield the samplers from the majority of the solids. Although this orientation will reduce the amount

of flow around the samplers, the samplers may perform better because of a minimal chance of damage or clogging from the suspended solids.

Vandalism

Vandalism is the greatest risk to the use of passive samplers in the field. Theft of the samplers is costly in terms of the hardware and the missed temporal component of the study. Fieldwork is often planned during a certain period of time related to seasons, biological cycles, or land-use activities. Vandalism also may result in the samplers being removed from the water and left on the land resulting in the loss of the temporal component of study but not the hardware costs.

Careful consideration is necessary to avoid deploying the canisters in easily accessible areas. Remember, if the study site is easy for the field crews, then it is easy for vandals to reach. Areas with boat traffic and fishing should also be avoided. Signage and markings are sometimes used with mixed results to identify the owner and purpose of the samplers. When public areas such as fishing access points and park land are unavoidable, deployments during the off-season and when schools are in session can reduce the chances of vandalism. Cooler weather and fewer hours of daylight can discourage outdoor activities. The best option is to hide and protect the deployment canisters as much as possible.

3c. Deployment Ideas

Methods for deploying passive samplers can vary greatly depending on the body of water to be sampled. The following section provides some suggestions based on successful methods used in other studies deploying passive samplers.

Lakes and Oceans

Deployment of passive samplers in large water bodies can be complicated by depth and the availability of structures for securing the passive samplers. Several options for securing passive samplers in lakes and oceans are given below.

Shore Deployment

Deployment canisters can be secured to a fixed point on the shore such as a tree, boulder, fence post, ground anchor, dock, using cables (figs. 3, 4). The canisters can be placed in position by wading or tossed out into the water. When tossing the canisters, precautions should be made that there are no underwater structures (rocks, trees) which will catch the canisters during the retrieval stage. If deploying from a dock, the canisters are generally suspended below the dock. Be aware that docks are common sources of wood preservatives and boat fuels which will be sampled by the SPMDs and POCIS. Docks may also be open to the public increasing the risk of vandalism.

Boat Deployment

Boats are often necessary to reach study sites on large bodies of water. Deployment canisters can be suspended off the bottom by attaching to piers, pilings, floating platforms, buoys, or other structures (fig. 5). Alternatively, canisters can be suspended from the bottom using a combination of floats and anchors. Divers or surface markers may be necessary to find and retrieve canisters not secured to the surface. If floats are used as markers in tidal areas, it is important to use a two-float system where a second float is attached with a long leader to the first float (fig. 5). This approach will aid in finding the samplers during a high tide where the first float may be underwater. Exhausts from boat motors can rapidly contaminate SPMDs; therefore it is necessary to shut off all motors before opening the shipping cans containing the passive samplers.

Other Deployment Considerations

Regardless of the deployment method, additional factors can affect the deployment and ultimately the data obtained.

Bottom Composition

If the canisters are placed on the lake bottom, it is important to know the condition of the upper sediment layer. If the bottom is soft or the area is susceptible to sedimentation, it is possible that the canisters may become buried during the deployment period. In these cases, it may be difficult to

Figure 3. Methods of attaching deployment canisters to shore and mid-channel of streams. Photographs by John Tertuliani, U.S. Geological Survey (top row), Wade Bryant, U.S. Geological Survey (bottom left and center), and Christopher Guy (U.S. Fish and Wildlife Service, bottom right).

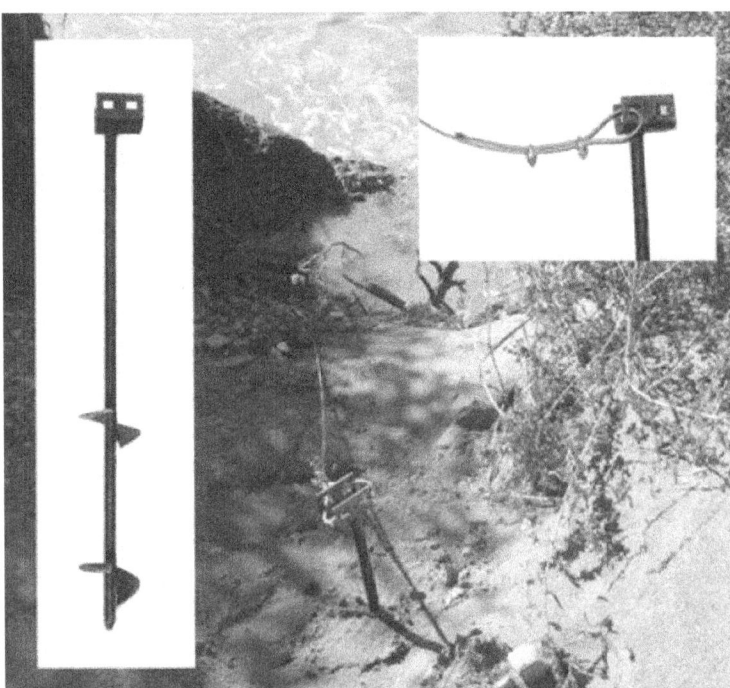

Figure 4.   Ground anchors (insets) can be useful for shore deployments when no other structures (trees, fence posts) are available for attachment purposes. Ground anchors are very strong, but may be difficult to impossible to use in rocky soils. Photographs by David Alvarez, U.S. Geological Survey.

Figure 5.   Deployment options in deep water suspending the canister off the bottom or for determining the vertical gradient of contaminant concentrations. Photographs by Kelly Smalling, U.S. Geological Survey (top left), David Alvarez, U.S. Geological Survey (top center, top right), Michael Rosen, U.S. Geological Survey (bottom left), and Jennifer Morace, U.S. Geological Survey (bottom center).

recover the canisters and more importantly, during burial, the samplers will be sampling the sediment and sediment pore water, not the water column. Estimates of water concentrations would not be representative of the water column. This may be avoided by placing the canisters on top of a block or other structure or by using a float to suspend the canister over an anchor (fig. 5).

Vertical Gradients

Depending on the depth of the water body, substantial gradients in the concentrations of contaminants can occur with depth. Seasonal differences in water temperature, density, and potential inputs such as effluent streams can all affect where in the water column the highest concentrations of contaminants may occur. To study this, samplers can be placed at various depths (fig. 5).

Photodegradation

Many organic contaminants such as PAHs, some PBDEs, some pharmaceuticals, and others are known to rapidly degrade upon exposure to sunlight. Chemicals sampled by SPMDs are susceptible to photodegradation as the polyethylene membrane of the SPMD is mostly transparent to UV radiation. The polyethersulfone membrane of the POCIS is opaque, therefore, photodegradation is not likely a problem. The deployment canisters offer minimal protection from sunlight as the holes which allow water to pass through also allow substantial amounts of light to enter. Although the rates of photolysis are much less in water than in air, substantial

photodegradation has been observed in clear water at depths of 20 feet (Garrison and others, 2005). Options to help reduce the potential for photodegradation of sampled chemicals include placing the samplers out of direct sunlight or adding some type of sunshield. Section 2b discussed the option of adding a photolysis marker to determine if photolysis may have been a problem. A simple sunshield used to protect SPMDs deployed in high elevation lakes is shown in figure 6. These sunshields were suspended over deployment canisters with floats blocking direct sunlight, however, as light is refracted in the water column, the SPMDs were still exposed to indirect sunlight. Note that sunshields, such as these, can only be used in calm water bodies, otherwise they may act a sail moving the canisters in the direction of waterflow.

Biological Growth

The development of a biofilm on the SPMD surface is a common occurrence and has been discussed in Section 2. In this section, biological growth refers to macro-organisms which may grow inside or outside of the deployment canister. On occasion, organisms such as crayfish and crustaceans may enter the deployment canister at a young stage and become trapped as they grow. In brackish and marine waters, a buildup of barnacles may occur reducing the availability of water to be sampled. Predicting when a buildup of organisms may occur can be difficult as most brackish and marine deployments have not proven to be a problem. On occasion, however, the entire deployment canister may fill with barnacles making it difficult to recover the passive samplers.

Figure 6. Sunshield used in calm waters to limit potential photodegradation of labile chemicals. Photographs by David Alvarez, U.S. Geological Survey.

Rivers and Streams

River and stream deployments have an advantage over still water bodies as moving water increases the volume of water sampled; however, the methods for deploying the samplers can vary depending on the size (width and depth) of the system. For deep rivers, many of the deployment options discussed for lakes and oceans can be used. Deployment in smaller systems may be easier than lakes and oceans but may be limited by depth (see discussion in Section 3b). Concerns regarding the bottom composition and photodegradation of some contaminants (as discussed in Section 3c) should also be addressed in river and stream deployments.

Wadable Streams

Common methods for deploying samplers in wadable streams include cabling the canisters to the shore or securing in the stream channel using fence posts or rebar.

Shore and Streambank Deployment

Canisters can be cabled to a fixed point on land (tree, boulder, fence post, ground anchor) or placed in the water (figs. 2, 3). Depending on the stream bottom conditions, the canisters can be directly placed on the bottom or suspended from a concrete block (fig. 6). Concrete blocks and anchors can be used to help keep the canisters from moving with flow. A canister secured to a stream bank is vulnerable to being pulled up and onto the bank during a high flow event. An increase in flow pulls the canister directly downstream from the anchor point, a canister weighted down with concrete block is susceptible to moving downstream with a moderate increase in flow, high flow can whip the canister back and forth at the end of the cable. When the water recedes, the canisters come to rest on the bank, out of the water. Inspecting canisters after moderate rainfall is something to consider in the study plan. Depending on the time of year and local climate, more than one inspection may be necessary to insure the canister(s) at each site have not been moved by high water.

Mid-Channel Deployment

Canisters can be secured and suspended in the water column mid-channel by securing to a fence post or rebar inserted into the streambed (fig. 2). A mid-channel deployment is preferred in streams that tend to have variable flow (flashy) after a rain event. A canister deployed mid-channel is less likely to be swept to the banks during a high-water event. They are also more likely to remain underwater during dry periods as the mid-channel is likely to be the deepest part of the stream.

Nonwadable Streams and Rivers

If the stream or river is not wadable, then the canister needs to be secured to a solid structure either on land or in the water. If the stream or river is accessible by boat, some of the lake and ocean deployment options may be feasible. Docks, breakwalls, piers, and bridges may provide a solid fixture to temporarily attach a cable for deployment (fig. 7).

Figure 7. Stream deployments with the deployment canisters cabled to the shore, sitting directly on the streambed, and anchored by concrete blocks. Photographs by David Alvarez, U.S. Geological Survey (top left), John Tertuliani, U.S. Geological Survey (top right, bottom left), and Christopher Guy, U.S. Fish and Wildlife Service, (bottom right).

3d. Hardware

Many options exist for the types of hardware that can be used for securing the canisters during field deployment. Strength and protection from vandalism must be considered when selecting materials. Stainless steel hardware is preferred for prolonged water exposure and is necessary in marine environments to prevent corrosion. Wire rope (also called aircraft cable) is recommended for securing the canisters. Wire rope (1/8 in and 3/16 in) can be handled easily in the field and has the strength to survive most adverse conditions. The 3/16 in wire rope is difficult to cut (discouraging vandalism), therefore, it is desirable to have it precut before going to the field. Saddle-style wire clips (also called cable locks) are used to secure the ends of the cable. The use of lock washers and lock nuts is recommended in combination with the clips for maximum security. Generally, two clips are sufficient to secure the ends of the cable. Chains can be used, but are heavy and may require additional tools not commonly needed in the field. Ropes are not recommended because of lower strength and increased risk of loss because of vandalism, such as being cut by a knife. Many cables, chains, and other metallic items are coated with an oil to prevent corrosion and used as a lubricant during manufacture. These oils could act as a potential interference or source of targeted contaminants (PAHs and hydrocarbons). The hardware should be thoroughly cleaned before use with a degreasing solution or other solvents. If detergents are used to clean hardware, they should be thoroughly rinsed with water and followed by a rinse with an organic solvent such as acetone or hexane to remove any residual surfactants.

Some additional options of securing canisters to cables for deployment are shown in figure 8. In some instances, large nylon cable ties can be used to secure canisters. Caution should be used as the cable ties have limited strength and can easily be cut by vandals and break from rubbing against other materials.

3e. Field Observations and Measurements

Information on the deployment sites can be useful in the estimation of ambient water concentrations and the final interpretation of the data. Descriptions of the deployment site and surrounding land use are important (photos provide excellent information). An example field log sheet is given in appendix 1.

General observations which can be useful include:

• Bottom conditions (soft, rocky)

• Water movement (calm, low, moderate, high flow)— flow measurements may be useful in the final data interpretation, but are not directly used for estimating chemical concentrations

• Water conditions (clear, murky, high suspended sediment levels, surface film present, algal growth)

• Weather/air quality during field work

• Water temperature

• Condition of the samplers when retrieved (buried, moved downstream, exposed)

Water temperature measured at the beginning and end of the deployment can be useful and may be necessary. Occasionally, average water temperatures can be determined from real-time temperature loggers at nearby USGS stream-gaging stations or by the use of " TidbiT" temperature loggers that can be attached to the deployment canisters. Other water properties such as pH, suspended solids concentrations, may be useful when discussing chemical speciation, distribution, and fate, but are generally not collected as part of a passive sampler study.

Figure 8. Different options for cabling canisters during deployment in water bodies. Photographs by David Alvarez, U.S. Geological Survey (top far left, top left center, bottom row), Michael Rosen, U.S. Geological Survey (top right center), and Randal Clark, U.S. Geological Survey (top far right).

3f. Shipping and Handling Precautions

Contamination

Since the SPMDs and POCIS are designed to sample low levels of organic contaminants, steps to minimize potential contamination are necessary. Samplers exposed to air for long periods (> 30 minutes) may concentrate significant amounts of air pollutants. The SPMD is extremely sensitive to fumes from engine exhausts, oils, tars, fuels, paints, solvents, cigarette smoke, fragrances (perfumes, deodorants, sunscreens). Prepare all deployment hardware on site before opening the passive sampler cans. Exposed skin or gloves should be clean and free of lotions, insect repellants, and powders. If the field personnel use such products, indicate the type, brand, and active ingredient (if known) on the field log sheets. Make sure that the surface of the water at the site is not coated with a film or floating oils or solvents. If a film exists, use an oar or other device to disperse the film before placing the samplers in the water.

Shipment and Storage

The passive samplers should be transported to the field in clean airtight metal cans on blue or wet ice. If wet ice is used, it should be placed in plastic zip bags to help prevent leaking which could result in the metal shipping cans rusting. It is important that the cans are not opened before use to prevent potential contamination from airborne chemicals. The cans containing the passive samplers should preferably be stored at < 0 °C or at a minimum, kept cool. Following deployment of the samplers, the shipping cans should be resealed to keep the cans clean for shipment of the samplers back to the laboratory after retrieval. During the deployment period, the empty cans with lids replaced should be kept in a safe place and the field blanks should be stored at < 0 °C. After retrieval, the samplers along with the field blanks, should be resealed in the original shipping cans, put on blue or wet ice and sent overnight to the laboratory. Dry ice should not be used as it can damage the passive samplers.

3g. Air Sampling

In addition to its utility for sampling water, the SPMD has also been demonstrated to be an effective sampling device for air (Huckins and others, 2006; Cranor and others, 2009). Most of the concerns regarding study design, handling, shipment, storage, and vandalism for water sampling studies are also valid for SPMD studies targeting air. Section 2c highlights the differences between field blanks and trip blanks. For air sampling studies, a trip blank is required to account for contamination during the transport to and from the study sites. Deployment of SPMDs in the air may require a protective deployment canister different from canisters used in the water. Photolysis of many potentially sampled chemicals is a primary concern; therefore, the deployment canister should keep the SPMD in complete darkness. The double-bowl canister design (fig. 9) is effective for protecting the SPMDs from light and although much of the airflow around the SPMD is reduced, there is still adequate air movement through the device via convection (Bartkow and others, 2007). Other devices described by Bartkow and others (2006) that were designed to reduce the amount of light were determined to be cumbersome and to allow reflected light inside the device, resulting in the photolysis of some PAHs sampled by the SPMD. In addition to the protection of the SPMD from light, the deployment canister should be weatherproof and easy to assemble in the field to minimize handling of the SPMDs.

It is recommended that the SPMDs used for air sampling be constructed containing PRCs and photolysis surrogates. Assuming the deployment canister negates potential photolysis problems and regulates airflow across the SPMD surface, there still will be diurnal temperature changes which may vary 10 to 20 °C in some climates. Uptake of a chemical into the SPMD has some dependence on temperature; therefore, PRCs will be important to correct for these site specific environmental changes.

The remaining concern with using passive samplers for air monitoring studies is the buildup of particulate matter on the sampler's surface. This particulate matter can contain a substantial portion of a target chemical, especially chemicals

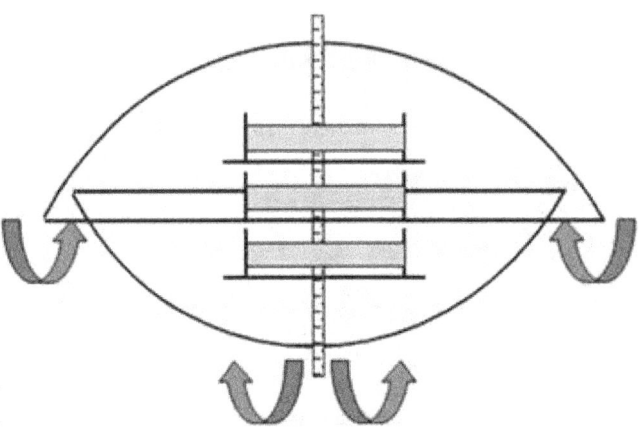

Figure 9. Double-bowl design for deploying semipermeable membrane devices (SPMDs) in air. Schematic on right shows three SPMDs contained inside the device (arrows show air movement direction around the lip of the bottom bowl and holes in the bottom of the device). Photograph and schematic by David Alvarez, U.S. Geological Survey.

with log octanol-air partition coefficients (K_{OA}s) greater than 9 (Bartkow and others, 2004). The buildup of particulates can complicate estimates of air concentrations and there is much debate on how to interpret the data. The SPMD is nonporous but can develop a sticky film on its surface through time which collects airborne particulate matter. During processing, the film and particulate matter can be removed, resulting in a measure of the airborne (not bound to particulates) chemicals. It is still unknown to what extent chemicals may partition from the particulate matter on the SPMD surface into the SPMD.

Polyurethane foam (PUF) disks are the most widely used passive samplers for air; however, the disks tend to accumulate particulate matter on the surface. This problem tends to be more substantial with PUF than the SPMD because the PUF surface is porous which allows the particulate matter to become incorporated into the PUF matrix. Currently, the best option is to use a protective deployment canister which reduces the exposure to particulate matter. The double-bowl design (fig. 9) has been reported to minimize, but not completely eliminate, exposure to particulate matter (Bartkow and others, 2007; Jaward and others, 2004).

4. Back at the Lab—Processing of the SPMD and POCIS

4a. Dialysis and Extraction

The first step in recovering sampled chemicals is to perform a dialysis procedure for SPMDs and an extraction procedure for the POCIS. These procedures have been well documented and are summarized below (Alvarez and others, 2004, 2008a; Huckins and others, 2006; Petty and others, 2000). At the time of this writing (2010), the dialysis of the SPMD is covered under U.S. and Canadian government patents and can only be performed by the commercial vendor (EST Labs) or by the USGS Columbia Environmental Research Center (Columbia, Mo., USA). Before performing this step, the user must check with the passive sampler supplier to determine if the patent protection is still in effect. Extraction of the POCIS is not covered under the U.S. government patent and, therefore, can be performed by any laboratory.

SPMD Dialysis

Each SPMD is individually removed from the storage container or support rack and immediately cleaned to remove any surficial particulate matter and biofilm. Cleaning is performed by scrubbing the SPMD surface with a gloved hand or soft toothbrush, quickly submerging in dilute acid to remove salts, rinsing with de-ionized water, and followed by a quick surface rinse with acetone then hexane. The cleaned SPMD is then placed in a contaminant-free glass container with an airtight lid containing a sufficient volume of hexane to cover the SPMD (fig. 10). The dialysis containers are then placed in an incubator at 18 °C for 18 to 24 hours. After this first dialysis period, the hexane is transferred into a separate container and a second portion of hexane is added to the container. This second dialysis period is performed for a minimum of 6 to 24 hours at which time the hexane from both dialysis periods is combined and the SPMD is discarded. The dialysis times may have to vary depending on the chemicals to be extracted. Generally, an 18-hour period followed by another 6 hours is sufficient for most chemicals; however, some chemicals such as the pyrethroids may require as many as three 24-hour periods in order to achieve adequate recovery. It should be noted that extended dialysis periods may result in an increased amount of coextracted matrix components in the sample.

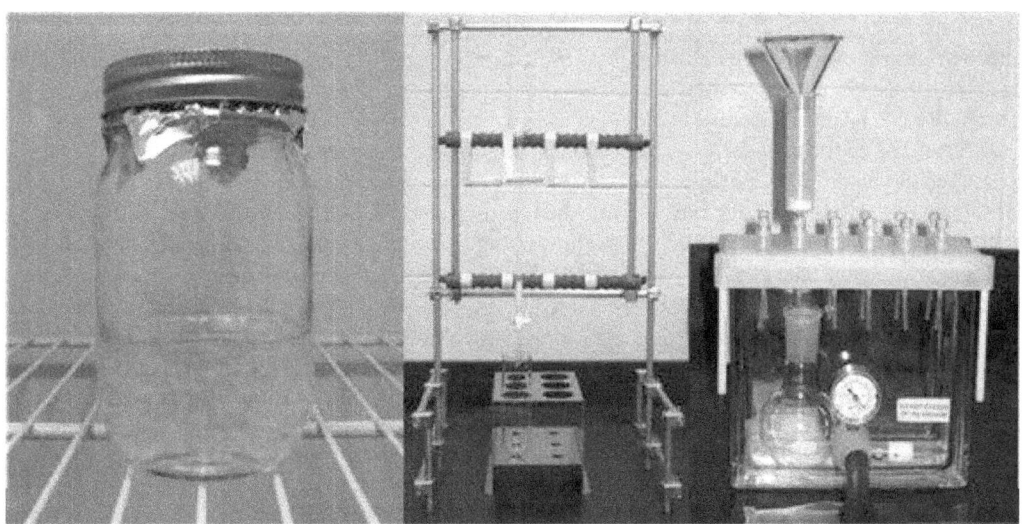

Figure 10.   A typical semipermeable membrane device (SPMD) dialysis (left), polar organic chemical integrative sampler (POCIS) extraction using glass chromatography columns (center), and POCIS extraction using solid phase extraction cartridges (right). Photographs by David Alvarez, U.S. Geological Survey.

POCIS Extraction

The extraction methods for the recovery of chemicals from the POCIS can vary depending on the equipment available at the laboratory. The two most common methods include the use of glass gravity-flow chromatography columns or empty solid phase extraction cartridges (fig. 10). Methods utilizing glass chromatography columns generally use a column (1 cm inner diameter x 30 cm long) fitted with a stopcock on one end and a solvent reservoir or funnel on the other. A glass wool plug is inserted next to the stopcock to keep the sorbent from washing through. An individual POCIS unit is carefully opened over a funnel on the top of the column and the sorbent is rinsed with a suitable solvent into the column taking care that all of the sorbent is transferred. Methanol or water is commonly used as the transfer solvent. If the transfer solvent is methanol, it is collected in a flask which will also receive the extraction solvent as it will contain a large percentage of the target chemicals. If water is used (generally not recommended for the gravity flow system as it is difficult to remove residual water) then the water would not be collected in the flask. A second glass wool plug is placed on top of the sorbent bed to prevent mixing as the extraction solvent is added.

There are three commonly used extraction solvent combinations used depending on the POCIS configuration (sorbent type) and the chemicals targeted. For the pesticide-POCIS configuration, 50 mL of a 1:1:8 (v:v:v) mixture of methanol :toluene:dichloromethane is used. With the pharmaceutical-POCIS configuration, 40 mL of methanol is used for most chemicals. If some of the targeted chemicals have a higher volatility, for example tetrachloroethylene, methanol should not be used; it requires more rigorous evaporation methods resulting in the loss of some volatile chemicals. An alternate method is to use 25 mL of 8:2 (v:v) dichloromethane methyl-*tert*-butyl ether (Alvarez and others, 2008b). Alternating the volumes and compositions of the extraction solvents can optimize recovery of targeted chemicals and should be determined before extracting any of the field deployed POCIS. This may require method development using fresh POCIS sorbents and mixtures of the targeted chemicals.

If the solid-phase extraction cartridge method is used, it is recommended that large empty cartridges with capacities of 15 mL or greater be used (fig. 10). In this method, a frit is placed in the bottom of a cartridge. The cartridge and frit are rinsed with the solvents to be used during the POCIS extraction and dried. The flow of solvent through the cartridge can be achieved by either gravity or using a vacuum manifold. Similar to the glass chromatography columns, the POCIS is opened over a funnel and a transfer solvent is used to rinse the sorbent into the cartridge. With this method, water is generally used to transfer the sorbent as the cartridge can then be dried by vacuum or high purity nitrogen to remove all traces of water before extraction. As described above, the extraction solvent used is dependent on the POCIS type and the suite of chemicals targeted.

Generally, only the solid sorbent inside the POCIS is extracted and the membranes are discarded. Studies have determined that POCIS membranes contain a small part of the sampled chemicals; however, in a 21 or more day study the amount of a chemical in the membrane was minimal compared to the concentration in the sorbent (Alvarez, 1999; Erhunse, 2010). Extracting targeted chemicals from the membrane uses a dialysis method similar to the dialysis of the SPMD and a suitable solvent such as methanol. Problems can arise when extracting chemicals from the membrane including potential chemical interferences coextracted from the membrane and the particulate matter embedded on the surface. Published sampling rates for the POCIS do not consider the membrane, therefore, any water concentration estimates based on data from the membrane and sorbent extraction would be an overestimation of the actual concentration.

4b. Quality Control

A thorough QC plan covers the field and laboratory components of the study. Section 2c outlines common QC sample types which are recommended to ensure the data are valid for the goals of the study. In addition to these recommended QC types, any QC procedures associated with the laboratory's routine GLP plan is desired. These GLP procedures often include development of standard operating procedures, instrument and analyst certification, documentation and maintenance of records, data reporting, evaluation, and review.

4c. Sample Composites and Splits

Section 2a introduced the idea of compositing SPMD or POCIS extracts into a single extract in order to increase the amount of a chemical in the final extract aiding in the detection at very low concentrations, such as dioxins. Compositing SPMD or POCIS extracts may also be desired when the sampling rate for a target chemical is very low or short exposure times were used resulting in minimal volumes of water extracted. Because of the small surface area of the POCIS, which is related to the amount of chemical sampled, it is a common practice to composite the extracts of two or more POCIS into a single extract in order to increase the amount of chemical present in the extract for detection.

Splitting of an extract between analyses can be useful when there are insufficient numbers of passive samplers to allow a single passive sampler to be dedicated for each analysis type or incompatible processing steps are needed for the targeted analyses. For example, a SPMD dialysate may be split equally between the processing steps for PAHs and those for chlorinated pesticides and PCBs. Careful consideration is important to ensure that the splitting of an extract will not reduce the measured concentration below the desired level (Section 2a, equation 1).

4d. Solvent Exchanges

Changing the solvent composition of extracts, regardless of the sample matrix, is a common practice in an analytical laboratory. Solvent exchanges can be performed by evaporation of a solvent and replacing it with a different solvent or by the partitioning of chemicals between two immiscible solvents. If volatile or semivolatile chemicals are to be targeted, it is important to not evaporate the initial solvent to dryness during the exchange process. The preferred method to remove the initial solvent while minimizing loss of targeted chemicals is to evaporate the solvent to near dryness, add the exchange solvent, and repeat these steps as necessary.

SPMD samples are generally in hexane or another nonpolar solvent following dialysis or subsequent processing. POCIS extracts are commonly in methanol, however, other solvent compositions may be used depending on the solvent(s) required in the initial extraction. When performing a solvent exchange, it is important to ensure the targeted chemicals are soluble in the final solvent.

4e. Cleanup and Fractionation

A literature survey will document environmental scientists often forgo a chemical specific sample cleanup and fractionation scheme, preferring instead to allow a mass spectrometer and deconvolution software to extract the signature of a targeted chemical out of the myriad of other coextracted chemicals. Nontargeted chemicals can be expected to be present at concentrations that are orders of magnitude greater than that of the targeted chemicals. This approach can cause problems as cosampled chemicals and matrix interferences may lead to misidentification of target chemicals, shifting chromatographic retention times, and ion suppression in mass spectrometric analyses (Chen and others, 2006; Jones-Lepp and others, 2004; 2009; Petrović and others, 2005).

Extracts from SPMDs and POCIS are generally considered to be "cleaner" than many other environmental matrices; however, there are specific matrix specific background chemicals that may occur. The SPMD dialysates may contain methyl oleate and oleic acid at levels varying with the type of triolein used in their construction. If an extended dialysis period is used, for example when trying to recover pyrethroids, some polyethylene waxes from the membrane may be solubilized. Elemental sulfur, a background contaminant not originating from the SPMD matrix but from the environment, is readily sampled by SPMDs and may be present as elevated levels. The presence of sulfur in environmental extracts is common, especially with sediments, and standard removal processes such as copper treatment or size exclusion chromatography can be used.

Matrix specific interferences are not as well defined for the POCIS as the sorbents are extensively cleaned before POCIS construction; however, polyethylene glycols, byproducts from the membrane polymerization process, have been reported in early studies using POCIS (Alvarez, 1999). Any other background interferences identified in SPMDs and POCIS are generally not associated with the passive sampler's matrix, but the result of contamination during sample processing. An adequate use of QC samples should account for these potential interferences.

Processing methods for the cleanup and analysis of PAHs, chlorinated pesticides, PCBs, dioxins and furans in SPMDs are well established (Alvarez and others, 2008a; Gale, 2007; Petty and others, 2000). Although tailored specifically for SPMDs, the methods are based on the processing and analysis of biota and sediment samples. As of 2010, processing methods for the POCIS are not as well defined. Cleanup methods for most of the common chemicals targeted by the POCIS (pharmaceuticals, hormones, surfactants) are lacking. A cleanup method for triazine herbicides and other current-use pesticides has been developed for use with the POCIS (Alvarez and others, 2008a).

Each laboratory will have a different cleanup and fractionation method for a specific chemical class determination. There is a vast assortment of published methods available in the peer-reviewed literature and the Internet. It is recommended to give priority to those methods with at least some cleanup capability to achieve the highest quality results.

4f. Storage and Shipment of Extracts

Following the extraction and any other processing of the samplers, SPMD and POCIS extracts will be in an organic solvent and should be stored according to the requirements of the targeted chemicals. Often, samples can be kept at room temperature in capped containers for short periods, days to weeks, while the processing continues. For extended periods, extracts are generally stored at <-20 °C. Some chemicals may be photosensitive and therefore should be kept in the dark. The final criterion for storage requirements is the stability of the specific chemicals to be targeted in the solvents and containers used. Shipment of extracts should follow the safety requirements of the laboratory. Generally, the extract is sealed under high purity nitrogen or argon in an amber glass ampoule and packaged according to the guidelines of the U.S. Department of Transportation for shipping minimal quantities of organic solvents.

5. I Have Data, Now What?

The analytical laboratory will provide the user chemical concentration data in a raw form that will require additional data processing in order to be useful. The data will usually be reported as nanograms or micrograms of a chemical per sample where the sample will be the passive sampler extract. In this simplest form, the data can be used to determine the presence or absence of a chemical at a site during the sampling period and it can qualitatively be used to rank sites based on the total mass of chemical(s) measured. For many of the chemicals sampled by the SPMD and POCIS, it is possible to estimate the time-weighted average concentration at the site during the sampling period. The following sections provide a discussion on how to convert the raw passive sampler data into average ambient water concentrations.

5a. Estimating Water Concentrations from SPMD Data

Using models previously developed (Huckins and others, 2006), the average water concentrations of selected chemicals can be estimated using PRC loss data, chemical sampling rates (when available), and amounts of chemicals sampled. The uptake of chemicals into passive samplers follows integrative (linear), curvilinear and equilibrium phases during the deployment/exposure period. Integrative uptake is the predominant phase for compounds with log K_{ow} values ≥ 5.0 and exposure periods of as much as one month in SPMDs. During the integrative uptake phase the ambient chemical concentration (C_w) is determined by

$$C_w = \frac{N}{R_s t} \tag{2}$$

where

N		is the amount of the chemical accumulated by the sampler (typically ng),
R_s		is the sampling rate (L/d), and
t		is the exposure time (d).

For SPMDs, regression models have been created which estimate a chemical's site specific R_s and its C_w based on the log K_{ow} of the chemical, the PRC's release rate constant (k_e) and SPMD-water partition coefficient (K_{sw}) (Huckins and others, 2006). A PRC's k_e is determined from the amount of PRC initially added to the SPMD (N_o) and the amount remaining (N) as shown in equation 3. The log K_{sw} is determined from a regression model of the PRC's log K_{ow} as shown in equation 4 where a_0 is the intercept determined to be -2.61 for PCBs, PAHs, nonpolar pesticides and -3.20 for moderately polar pesticides. The $R_{s\text{-}PRC}$ can then be calculated as shown in equation 5 where V_s is the volume of the SPMD (in L or mL).

$$k_e \frac{\left[\ln\left(\dfrac{N}{N_o}\right)\right]}{t} \tag{3}$$

$$\log K_{sw} = a_o + 2.321 \log K_{ow} - 0.1618\,(\log K_{ow})^2 \tag{4}$$

$$R_{s-PRC} = V_s K_{sw} k_e \tag{5}$$

The extrapolation of C_w from measured values of N requires knowledge of a chemical's site-specific sampling rate (R_{si}) which is determined from a third-order polynomial (eq. 6) where $\alpha_{(i/PRC)}$ is the compound-specific effect on the sampling rate and the relation between the $R_{s\text{-}PRC}$ and R_{si} (eq. 7).

$$\log\alpha_{(i/PRC)} = 0.0130\,\log K_{ow}^{\,3} - 0.3173\,\log K_{ow}^{\,2} + 2.244\,\log K_{ow} \tag{6}$$

$$R_{si} = R_{s\,PRC}\left(\frac{\alpha_i}{\alpha_{PRC}}\right) \tag{7}$$

The C_w of a chemical in the water can then be calculated by

$$C_w = \frac{N}{\left[V_s K_{sw}\left[1 - \exp\left(\dfrac{-R_s t}{V_s K_{sw}}\right)\right]\right]} \tag{8}$$

Huckins and others (2006) describe in detail the derivations and the theoretical aspects of these and other models. To simplify the process of performing these calculations, a set of Microsoft Excel spreadsheets were created to calculate ambient water concentrations from site specific data (Alvarez, 2010a; 2010b). Two versions of the spreadsheet calculator can be downloaded from the USGS Passive Sampling group's website *http://www.cerc.usgs.gov/Branches.aspx?BranchId=8.* Version 5.1 (2010) uses the models described above with data from one or more PRCs (fig. 11). The ability to use multiple PRCs reduces the variability present in a single measurement by using an average value to determine the R_{si}. Version 4.1 (2010) estimates water concentrations from SPMD data when PRCs were not used (fig. 12). In this case, the number of chemicals which can be determined is limited to those which have experimentally-derived R_s values. This calculator uses early uptake models where each phase of the chemical uptake curve (integrative, curvilinear, and equilibrium) is described by separate equations. Based on the K_{SPMD} and the deployment time, a theoretical half-time ($t_{1/2}$) is calculated and is used to determine which model should be used to estimate the water concentration for a specific chemical (Huckins and others, 2002b). In cases where PRCs were not used and R_s data for a specific chemical is not available, the result should be reported as mass of chemical sampled per SPMD (ng/SPMD or µg/SPMD). Results expressed in this format are qualitative, as the actual water concentration is not known; however, with regards to the detection limit, presence or absence is defined, and the date can be useful in determining the relative amounts of a chemical present at each site (ranking of sites). Users should refer to the above website for the latest updates to these calculators.

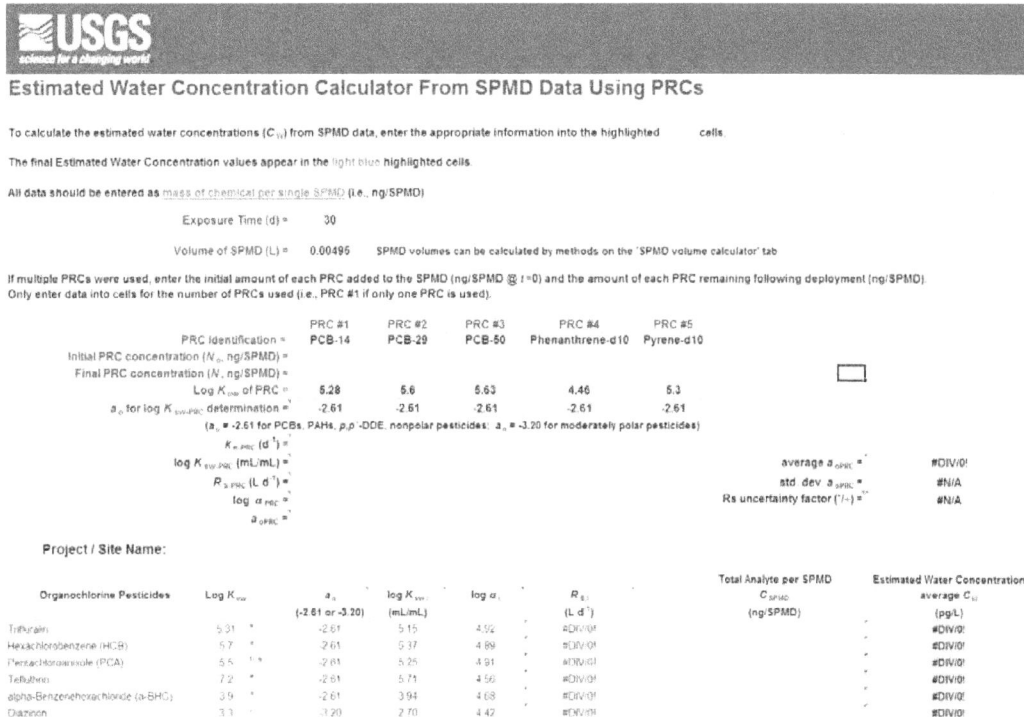

Figure 11. Downloadable spreadsheet to estimate water concentrations from a semipermeable membrane device (SPMD) data when using one or more performance reference compounds (PRCs).

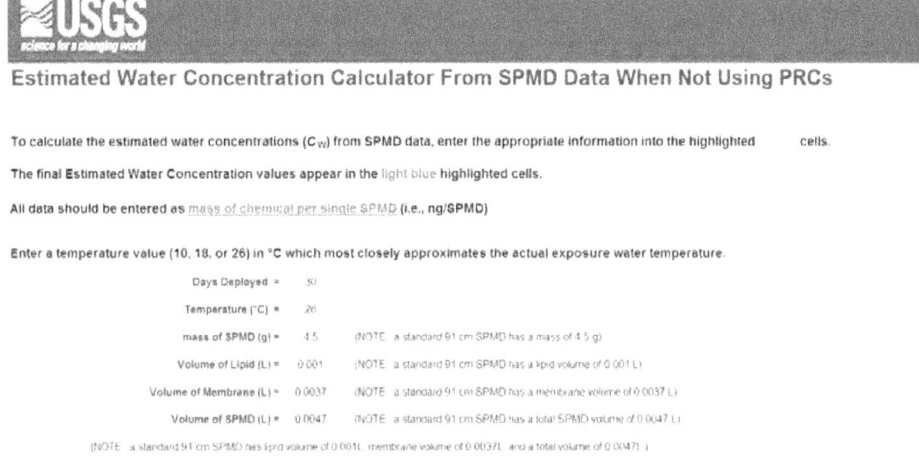

Figure 12. Downloadable spreadsheet to estimate water concentrations from a semipermeable membrane device (SPMD) data when performance reference compounds (PRCs) were not used.

These spreadsheet calculators are designed to be simple for the casual user, as well as provide additional data related to the variables and models used for the experienced user. In the simplest terms, a user enters data in the yellow highlighted cells and the estimated water concentrations are shown in the blue highlighted cells. Specific instructions are given on the "Instructions" tab at the bottom of each spreadsheet. A common mistake when using the calculators is not converting all data into the units of nanograms of chemical per a single SPMD (ng/SPMD). The conversion should include any sample composites or splits that were performed.

5b. Estimating Water Concentrations from POCIS Data

Estimating water concentrations from POCIS data is currently limited by the availability of experimentally-derived R_s data. Available R_s data are summarized and incorporated into a downloadable spreadsheet on the USGS Passive Sampling group's website *http//www.cerc.usgs.gov/Branches. aspx?BranchId=8*. In cases where R_s data for a specific chemical is not available, the result should be reported as mass of chemical sampled per POCIS (ng/POCIS or μg/POCIS). Results expressed in this format are qualitative as the actual water concentration is not known; however, they should be used to indicate the presence or absence of a chemical. With regards to the detection limit, this information can be useful in determining the relative amounts of a chemical present at each site (ranking of sites).

Previous data indicates that many chemicals of interest sampled by the POCIS remain in the integrative phase of sampling for at least 56 d (Alvarez and others, 2004; 2007), therefore, the use of the integrative uptake model (eq. 2) for the calculation of ambient water concentrations is justified. As new data and estimation models are made available, the information will be posted to the above website. To minimize calculation mistakes involving the compositing or splitting of POCIS extracts, it is recommended to always convert results into units of mass of chemical per single POCIS (ng/POCIS) before performing any comparisons or estimation of water concentrations.

5c. Estimating Air Concentrations from SPMD Data

The use of SPMDs to determine concentrations of semivolatile organic chemicals in the air was first published by Petty and others (1993). Following that initial work, there has been limited research investigating the use of SPMDs as passive air samplers (Cranor and others, 2009). Cranor and others (2009) present the most extensive work to date investigating the uptake rate constants and partition coefficients for a range of organic chemicals in air for SPMDs. This work included a series of PAHs, chlorinated pesticides, PBDEs,

phthalate esters, organophosphate and organosulfur pesticides, and synthetic pyrethroids. Uptake rate constants (k_{ua}) for 50 test chemicals along with models to estimate ambient air concentrations were presented in that work. Briefly, the air concentration of a chemical (C_a) can be determined as shown in equation 9.

$$C_a = \frac{C_s}{k_{ua}t} \qquad (9)$$

The concentration of a chemical in the SPMD is given by C_s in units of nanogram (or microgram) of chemical per g of SPMD, the k_{ua} is in units of $m^3g^{-1}d^{-1}$, and the deployment time, t, is in days.

5d. Interpreting Quality Control Information

Quality control data (blanks and spikes) provide validity to the measurements obtained from the chemical or biological analyses. Unfortunately, there are no universally accepted methods for how to use the QC results. How QC data are interpreted varies among agencies, laboratories, and at times, research groups. The following text suggests how various types of QC can be used; however, the user should check and follow the prescribed QC policies of their organization.

Background Correction

There are three main schools of thought regarding how to address the presence of a targeted chemical in a blank sample (fabrication, field, trip, or laboratory). In general, the most common method is to subtract the concentration of the chemical in the blank from the concentration of the chemical in the environmental (exposed) sample. The second method compared the concentration of a chemical in a blank to the concentration in an environmental sample, but is not subtracted. If the concentration of a chemical in an environmental sample is a defined factor (3, 5, or possibly 10 times) greater than the chemical concentration in the blank, then the environmental data can be used. If the chemical concentration in the environmental sample is less than this defined uncertainty factor, then the environmental data is censored accordingly. The third method uses the blank chemical concentration to establish the method detection limit (MDL) and method quantitation limit (MQL). An average MDL and MQL for a chemical can be determined by taking the average concentration measured in all the field blanks. Site specific MDL and MQL values can be determined using the appropriate field blank if blank data are available at each site. The third method is similar to the second method in that the data is censored by a specific factor determined by the blanks; however, the blanks are not subtracted from the value. Concentrations of chemicals in the blanks are accounted for by raising the MDL and MQL with respect to the average concentration and variability in the blanks.

Recovery Correction

Depending on the requirements of the study or organization, environmental data may be recovery corrected based on results from the matrix samples that were fortified with the targeted chemicals and analyzed concurrently with the field samples. This correction involves dividing the decimal equivalent of the percent recovery into the uncorrected field value. Keith (1991) recommends that environmental values should not be recovery corrected and all supporting QC data should be provided. If recovery corrected data is warranted, the recovery data should be supplied so that the original uncorrected values can be interpreted as needed.

The increased use of mass spectrometry in the environmental analytical laboratory has made available the technique of isotopic dilution, a process of adding stable-labeled (deuterium or ^{13}C) isotopes of the targeted chemicals into a sample before or immediately after extraction. Since the isotopes are affected by the same sample processing and potential interferences, they can serve as an internal recovery monitor. Although this method is attractive, it is used infrequently because of the expense and often lack of available isotopically-labeled standards and instrumental methods.

Determination of Detection and Quantitation Limits

As with other QC parameters, the opinions and methods on how to calculate MDL and MQL values widely vary. The terms reporting limit (RL), laboratory reporting limit (LRL), minimum reporting level (MRL), long-term method detection level (LT-MDL), instrumental detection limit (IDL), limit of detection (LOD), limit of quantitation (LOQ), practical quantitation limit (PQL), and target detection limit (TDL) vary as well among the different laboratories and organizations. Some of these terms can be used interchangeably, but care should be taken to recognize the subtle differences between terms. Definition of these terms is beyond the scope of this document. Users should consult with their laboratory manager or water quality specialist to understand how the limits applied to their data were generated and how they affect the final data set.

MDL or MQL values provided by the laboratory in units of ng/SPMD or ng/POCIS are not informative to the general public. A resource manager would need to know what was the lowest possible concentration attainable in terms of an average chemical concentration in the water or air. To answer this question, the MDL and MQL values are entered into the downloadable water concentration spreadsheet calculator (Sections 5a and 5b). First, the MDL and MQL values need to be determined. One example is to determine the average measured chemical concentrations from the field blanks as well as the standard deviation of those measurements for a particular chemical. Using the MDL and MQL definitions adopted from Keith (1991), the MDL is the mean of the blank measurements for a single chemical plus three times the standard deviation.

The MQL is the mean of the blank measurements plus ten times the standard deviation. Other methods to determine MDL and MQL could be used as well. In cases where PRCs were used, the simplest method is to take the average PRC results from all sites and use those numbers in the MDL and MQL estimations in the spreadsheet calculator. This adjusts the MDLs and MQLs for the average site-specific environmental factors. A more rigorous approach would be to use the site-specific PRC results which would provide a specific MDL and MQL set of values for each site. If PRCs were not used or the data originated from the POCIS, the MDL and MQL values would be used without any correction for environmental conditions.

5e. Bioassay Results

The use of SPMD and POCIS in bioindicator tests, classic toxicity tests, and whole organism dosing/exposure studies is commonplace. Typically, the generation of results will follow the established procedures of that test. In cases where the result of the test is given as a relative amount of a chemical per sampler, it may be possible to estimate the result in terms of a water concentration using the models presented above. For example, the YES assay generally gives a result as an equivalent concentration of the natural hormone 17β-estradiol which would give the observed estrogenic response. In the case of a POCIS extract, a result of X nanograms of 17β-estradiol per POCIS could be converted into a water concentration of Y nanograms of 17β-estradiol per liter using equation 2, the R_s for 17β-estradiol, and the number of days the POCIS was deployed.

6. Wrap Up and Review

The use of passive samplers in environmental monitoring studies is widely becoming an accepted practice. Passive samplers are easy to use and potentially cost effective means to sample chemicals occurring at extremely low concentrations in the environment. The information presented is meant to guide users towards a successful passive sampling study. Many of the topics covered are specific to SPMDs and POCIS, but are applicable to passive samplers in general as well as most environmental studies regardless of the sample matrix collected. A successful study includes a thorough plan, detailing what will be measured, how the sample will be collected, by whom, how the sample will be analyzed, ensuring adequate QC is used to validate the results, and how the results will be interpreted for the client. The guidelines in this document are intended to help new and experienced users of passive samplers think about the needs and potential pitfalls surrounding their study. The author acknowledges this document does not cover all potential situations one may encounter during a study, but can increase awareness needed to plan and complete an effective study using passive samplers.

7. References Cited

Alvarez, D.A., 1999, Development of an integrative sampling device for hydrophilic organic contaminants in aquatic environments: University of Missouri-Columbia, Ph.D. dissertation.

Alvarez, D.A., Petty, J.D., Huckins, J.N, Jones-Lepp, T.L., Getting, D.T., Goddard, J.P., and Manahan, S.E., 2004, Development of a passive, in situ, integrative sampler for hydrophilic organic contaminants in aquatic environments: Environmental Toxicology and Chemistry, v. 23, p. 1,640–1,648.

Alvarez, D.A., Huckins, J.N., Petty, J.D., Jones-Lepp, T.L., Stuer-Lauridsen, F., Getting, D.T., Goddard, J.P., and Gravell, A., 200 , Tool for monitoring hydrophilic contaminants in water: polar organic chemical integrative sampler (POCIS) *in* Greenwood, R., Mills, G., Vrana, B., eds., Passive Sampling Techniques: Comprehensive Analytical Chemistry, v. 48, Elsevier, p. 171–197.

Alvarez, D.A., Cranor, W.L., Perkins, S.D., Clark, R.C., and Smith, S.B., 2008a, Chemical and toxicological assessment of organic contaminants in surface water using passive samplers: Journal of Environmental Quality, v. 37, p. 1,024–1,033.

Alvarez, D.A., Cranor, W.L., Perkins, S.D., Schroeder, V.L., Werner, S., Furlong, E.T., and Holmes, J., 2008b, Investigation of organic chemicals potentially responsible for mortality and intersex in fish of the North Fork of the Shenandoah River, Virginia, during spring of 2007: U.S. Geological Survey Open-File Report 2008–1093, 16 p. (Also available at *http://pubs.usgs.gov/of/2008/1093/*).

Alvarez, D.A., 2010a, Estimated Water Concentration Calculator from SPMD Data Using Multiple PRCs: Version 5.1. Microsoft Excel spreadsheet. *http://www.cerc.usgs.gov/Branches.aspx?BranchId=8*.

Alvarez, D.A., 2010b, Estimated Water Concentration Calculator from SPMD Data When Not Using PRCs: Version 4.1. Microsoft Excel spreadsheet. *http://www.cerc.usgs.gov/Branches.aspx?BranchId=8*.

Arditsoglou, A., and Voutsa, D., 2008, Passive sampling of selected endocrine disrupting compounds using polar organic chemical integrative samplers: Environmental Pollution, v. 156, p. 316–324.

Bartkow, M.E., Huckins, J.N., and Müller, J.F., 2004, Field-based evaluation of semipermeable membrane devices (SPMDs) as passive air samplers of polyaromatic hydrocarbons (PAHs): Atmospheric Environment, v. 38, p. 5,983–5,990.

Bartkow, M.E., Kennedy, K.E., Huckins, J.N., Holling, N., Komarova, T., and Müller, J.F., 2006, Photodegradation of polyaromatic hydrocarbons in passive air samplers: Field testing different deployment chambers: Environmental Pollution, v. 144, p. 371–376.

Bartkow, M.E., Orazio, C.E., Gouin, T., Huckins, J.N., and Müller, J.F., 2007, Towards quantitative monitoring of semivolatile organic compounds using passive air samplers in Greenwood, R., Mills, G., and Vrana, B., eds., Passive Sampling Techniques: Comprehensive Analytical Chemistry, v. 48, Elsevier, p. 125–137.

Buchberger, W.W., 2007, Novel analytical procedures for screening of drug residues in water, waste water, sediment and sludge: Analytical Chimica Acta, v. 593, p. 129–139.

Bueno, M.J.M., Hernando, M.D., Agüera, A., and Fernández-Alba, A.R., 2009, Application of passive sampling devices for screening micro-pollutants in marine aquaculture using LC-MS/MS: Talanta, v. 77, p. 1,518–1,527.

Chen, J., Lichwa, J., Snehota, M., Mohanty, S., and Ray, C., 2006, Determination of hormones and non-ionic surfactant degradation products in small-volume aqueous samples from soil columns using LC-ESI-MS-MS and GC-MS: Chromatographia, v. 64, p. 413–418.

Cranor, W.L., Alvarez, D.A., Huckins, J.N., and Petty, J.D., 2009, Uptake rate constants and partition coefficients for vapor phase organic chemicals using semipermeable membrane devices (SPMDs): Atmospheric Environment, v. 43, p. 3,211–3,219.

Erhunse, A., 2010, An assessment of perfluoro alkyl substances bioavailability and policy implication for water quality and biota in the Lower Apalachicola River and Estuary: Tallahassee, Fl, University of Florida A&M, Ph.D. dissertation, 250 p.

European Commission, 2009, Common Implementation Strategy for the Water Framework Directive (200/60/EC): Guidance Document No. 19, Technical Report 2009-025, 132 p.

Gale, R.W., 2007, Estimation of freely-dissolved concentrations of polychlorinated biphenyls, 2,3,7,8-substituted congeners and homologs of polychlorinated dibenzo-*p*-dioxins and dibenzofurans in water for development of total maximum daily loadings for the Bluestone River watershed, Virginia and West Virginia: U.S. Geological Survey Open-File Report 2007–1272, 27 p.

Garrison, V.H., Orazio, C.E., Alvarez, D.A., Aeby, G., Carroll, J., and Taylor, M., 2005, Anthropogenic organic contaminants on coral reefs – global atmospheric deposition or local sources? U.S. Geological Survey Open-File Report 2005–1400, 24 p.

Goodbred, S.L., Bryant, W.L., Rosen, M.R., Alvarez, D.A., and Spencer, T., 2009, How useful are the "other" semipermeable membrane devices (SPMDs); the mini-unit (15.2 cm long)?: Science of the Total Environment, v. 407, p. 4,149–4,156.

Harman, C., Bøyum, O., Tollefsen K.E., Thomas, K.V., and Grung, M., 2008, Uptake of some selected aquatic pollutants in semipermeable membrane devices (SPMDs) and the polar organic chemical integrative sampler (POCIS): Journal of Environmental Monitoring, v. 10, p. 239–247.

Huckins, J.N., Lebo, J.A., Tubergen, M.W., Manuweera, G.K., Gibson, V.L., and Petty, J.D., 1992, Binary concentration and recovery process: U.S. Patent 5,098,573, March 24, 1992.

Huckins, J.N., Petty, J.D., Zajicek, J.L., and Gibson, V.L., 1995, Device for the removal and concentration of organic compounds from the atmosphere: U.S. Patent 5,395,426, March 7, 1995.

Huckins, J.N., Lebo, J.A., Tubergen, M.W., Manuweera, G.K., Gibson, V.L., and Petty, J.D., 1996, Binary concentration and recovery process: Canadian Patent 2,037,320, December 17, 1996.

Huckins, J.N., Petty, J.D., Lebo, J.A., Almeida, F.V., Booij, K., Alvarez, D.A., Cranor, W.L., Clark, R.C., and Mogensen, B.B., 2002a, Development of the permeability/performance reference compound approach for in situ calibration of semipermeable membrane devices: Environmental Science and Technology, v. 36, p. 85–91.

Huckins, J.N., Petty, J.D., Prest, H.F., Clark, R.C., Alvarez, D.A., Orazio, C.E., Lebo, J.A., Cranor, W.L., and Johnson, B.T., 2002b, A guide for the use of semipermeable membrane devices (SPMDs) as samplers of waterborne hydrophobic organic contaminants: Report for the American Petroleum Institute (API), API publication number 4690; API, Washington, DC.

Huckins, J.N., Petty, J.D., and Booij, K., 2006, Monitors of organic chemicals in the environment - semipermeable membrane devices: Springer, New York.

Jaward, F.M., Farrar, N.J., Harner, T., Sweetman, A.J., and Jones, K.C., 2004, Passive air sampling of polycyclic aromatic hydrocarbons and polychlorinated naphthalenes across Europe: Environmental Toxicology and Chemistry, v. 23, p. 1,355-1,364.

Jones-Lepp, T.L., Alvarez, D.A., Petty, J.D., and Huckins, J.N., 2004, Polar organic chemical integrative sampling (POCIS) and LC-ES/ITMS for assessing selected prescription and illicit drugs treated sewage effluent: Archives of Environmental Contamination and Toxicology, v. 47, p. 427–439.

Jones-Lepp, T.L., Alvarez, D.A., Englert, B., and Batt, A.L., 2009, Pharmaceuticals and hormones in the environment *in* Encyclopedia of Analytical Chemistry, Environment: Water and Waste, Meyers RA, Ed., John Wiley & Sons, Ltd., published online September 15, 2009. *http://onlinelibrary.wiley.com/doi/10.1002/9780470027318.a9059/abstract.*

Keith, L., 1991, Environmental sampling and analysis: a practical guide: Boca Raton, Fl, CRC Press, 143 p.

Lebo, J.A., Almeida, F.V., Cranor, W.L., Petty, J.D., Huckins, J.N., Rastall, A., Alvarez, D.A., Mogensen, B.B., and Johnson, B.T., 2004, Purification of triolein for use in semipermeable membrane devices (SPMDs): Chemosphere, v. 54, p. 1,217–1,224.

Li, H., Helm, P.A., and Metcalfe, C.D., 2010, Sampling in the Great Lakes for pharmaceuticals, personal care products, and endocrine-disrupting substances using the passive polar organic chemical integrative sampler: Environmental Toxicology and Chemistry, v. 29, p. 751–762.

MacLeod, S.L., McClure, E.L., and Wong, C.S., 2007, Laboratory calibration and field deployment of the polar organic chemical integrative sampler for pharmaceuticals and personal care products in wastewater and surface water: Environmental Toxicology and Chemistry, v. 26, p. 2,517–2,529.

Mazzella, N., Dubernet, J.F., and Delmas, F., 2007, Determination of kinetic and equilibrium regimes in the operation of polar organic chemical integrative samplers. Application to the passive sampling of the polar herbicides in aquatic environments: Journal of Chromatography A, v. 1,154, p. 42–51.

Mazzella, N., Lissalde, S., Moreira, S., Delmas, F., Mazellier, P., and Huckins, J.N., 2010, Evaluation of the use of performance reference compounds in an Oasis-HLB adsorbent based passive sampler for improving water concentration estimates of polar herbicides in freshwater: Environmental Science and Technology, v. 44, p. 1,713–1,719.

Pacáková, V., Loukotková, L., Bosáková, Z., and Štulik, K, 2009, Analysis of estrogens as environmental pollutants-A review: Journal of Separation Science, v. 32, p. 867–882.

Petrović, M., Hernando, M.D., Diaz-Cruz, M.S., and Barceló, D., 2005, Liquid chromatography-tandem mass spectrometry for the analysis of pharmaceutical residues in environmental samples: a review: Journal of Chromatography A, v. 1,067, p. 1–14.

Petty, J.D., Huckins, J.N., and Zajicek, J.L., 1993, Application of semipermeable membrane devices (SPMDs) as passive air samplers: Chemosphere, v. 27, p. 1,609–1,624.

Petty, J.D., Orazio, C.E., Huckins, J.N., Gale, R.W., Lebo, J.A., Meadows, J.C., Echols, K.R., and Cranor, W.L., 2000, Considerations involved with the use of semipermeable membrane devices for monitoring environmental contaminants: Journal of Chromatography A, v. 879, p. 83–95.

Petty, J.D., Huckins, J.N., and Alvarez, D.A., 2002, Device for sequestration and concentration of polar organic chemicals from water: U.S. Patent 6,478,961B2, November 12, 2002.

Petty, J.D., Huckins, J.N., Alvarez, D.A., Brumbaugh, W.G., Cranor, W.L., Gale, R.W., Rastall, A.C., Jones-Lepp, T.L., Leiker, T.J., Rostad, C.E., and Furlong, E.T., 2004, A holistic passive integrative sampling approach for assessing the presence and potential impacts of waterborne environmental contaminants: Chemosphere, v. 54, p. 695–705.

Rastall, A.C., Neziri, A., Vukonvic, Z., Jung, C., Mijovic, S., Hollert, H., Nikcevic, S., and Erdinger, L., 2004, The identification of readily bioavailable pollutants in Lake Shkodra/Skadar using semipermeable membrane devices (SPMDs), bioassays and chemical analysis: Environmental Science and Pollution Research, v. 11, p. 240–253.

Reemtsma, T., and Quintana, J.B., 2006, Analytical methods for polar pollutants, *in* Reemtsma, T., and Jekel, M., eds., Organic Pollutants in the Water Cycle, Weinheim, Wiley-VCH, p. 1–40.

Vermeirssen, E.L.M., Körner, O., Schönenberger, R., Suter, M.J.F., and Burkhardt-Holm, P., 2005, Characterization of environmental estrogens in river water using a three pronged approach: active and passive water sampling and the analysis of accumulated estrogens in the bile of caged fish: Environmental Science & Technology, v. 39, p. 8,191–8,198.

Waters Corporation, 2008, Oasis Applications Notebook, *http://www.waters.com/waters/library.htm?cid-513209&lid=1528415.*

Glossary and Abbreviation List

$\alpha_{(i\text{-}PRC)}$	compound-specific effect on the sampling rate
C_a	ambient chemical concentration in the air
C_c	environmental chemical concentration of concern
CERC	Columbia Environmental Research Center
C_w	ambient chemical concentration in the water
EST	Environmental Sampling Technologies, St. Joseph, MO USA
E_t	fraction of the total sample extract injected into the analytical instrument
GLP	good laboratory practices
ID	internal diameter
IDL	instrumental detection limit
k_e	chemical release (elimination) rate
K_{oa}	octanol-air partition coefficient
K_{ow}	octanol-water partition coefficient
K_{SPMD}	SPMD partition coefficient
K_{sw}	SPMD-water partition coefficient
k_{ua}	chemical uptake rate constant in air
LC/MS	liquid chromatography\mass spectrometry
LOD	limit of detection
LOQ	limit of quantitation
LRL	laboratory reporting limit
LT-MDL	long-term method detection level
MDL	method detection limit
MQL	method quantitation limit
MRL	minimum reporting limit
n	number of passive samplers composited into a single sample
N	amount of the chemical accumulated by the sampler
N_o	amount of PRC initially added to the SPMD
PAH	polycyclic aromatic hydrocarbon
PCB	polychlorinated biphenyl
PBDE	polybrominated diphenyl ether
POCIS	polar organic chemical integrative sampler
PQL	practical quantitation limit
P_r	overall method recovery for a target chemical
PRC	performance reference compound
PUF	polyurethane foam
QC	quality control
RL	reporting limit
R_s	sampling rate (liters per day)
R_{si}	a chemical's site-specific sampling rate
$R_{s\text{-}prc}$	sampling rate of the PRC

SPME	solid phase microextraction
t	time in days
$t_{1/2}$	theoretical half-time
TDL	target detection limit
UVA/UVB	ultraviolet radiation in the A and B band
V_i	volume of sample injected into the analytical instrument
V_s	volume of the SPMD
YES	yeast estrogen screen

Appendix

FIELD DATA SHEET – PASSIVE SAMPLERS

	Deployment	Retrieval
Location		
Date		
Time		
Field personnel		
Types/numbers of canister(s)		
Depth of canister(s)		
Water temperature		
Water conditions (clear, murky, high suspended solids, floating debris)		
Bottom conditions (if sampler is on bottom)		
Flow conditions (calm, moderate, rapid flow)		
Field blank used	Yes No #	Yes No #
Comments		

Field Equipment Checklist

- Ice chest/cooler for transporting the passive samplers to/from the field

- Deployment canister(s) in sealed metal cans

- Field/trip blank(s)

- Cold ice packs (blue ice or wet ice sealed in plastic bags)

- Tools

- Paint can opener

- Assorted tools (wrenches, pliers, cutters, saws)

- Thermometer

- Deployment hardware

- Cable, cable clamps

- Tie-down anchors, rebar

- Floats

- Signage, markings (personal choice because of potential vandalism)

- Safety

- Gloves (latex or other)

- Personal Floating Device

- Field notes / Data collection

- Field log book/sheets

- Pens, markers

- Digital camera

- GPS maps

- Other

www.ingramcontent.com/pod-product-compliance
Lightning Source LLC
Chambersburg PA
CBHW081411170526
45166CB00010B/3299